ELECTRONIC HARASSMENT

By Nolan Wilson

DESERT PUBLICATIONS

CORNVILLE, ARIZONA 86325

ELECTRONIC HARASSMENT

By Nolan Wilson

Illustrations by Jordan L. Cooper

ISBN: 0-87947-214-6

DESERT PUBLICATIONS
Cornville, Arizona 86325

Table Of Contents

Introduction

As the title emplies, this book covers a very important, but until now overlooked, facet in the science of harassment. When Desert Publications first addressed the subject of what is rightly entitled "Underdog Justice", an entirely new approach to guerrilla warfare was discovered. The dictionary defines the word "guerrilla" as meaning "little war" and the classic application throughout the ages is where a small but dedicated group engages a larger and better equipped enemy, usually with much success.

Techniques as presented in the previous *Techniques of Harassment* series reduced guerrilla warfare to its lowest common denominator to enable one individual to become a virtual army against his enemy. This book, as does the previously published series, makes no judgment as to the moral or legal ramifications involved. This judgment is the responsibility of the person who decides to initiate a harassment campaign against someone.

Almost everyone has had at least one experience in their life where they were wronged, many times severely, by some individual or group (company, government entity, etc.) and were seemingly helpless in defending themselves or at obtaining any form of justice. There seemed to be few places to turn for help until "Underdog Justice" became a reality.

In conventional warfare the wise opponent is one who avails himself of state of the art technology. Why should the individual underdog have less? In this book we will show many ways that the science of electronics can be used to great advantage. A knowledge of electronics is desirable, but not necessary. Some techniques may require rather unique electronic equipment while others are quite simple. Some of the things described are of a prankster nature with little if any harm done to the "target", while others could be devastating.

This book is presented in "scenario" style with each describing a complete technique. Most of what you are about to read has been successfully done by someone and documented by the author. Names have been changed to protect both the innocent and the guilty. You can be assured that the methods described will work if followed correctly.

Basics

While electronics is a far-reaching tool, we want to correct any false assumptions you may have. While appearing magical to the uninitiated, electronics is an exact science that follows definite physical laws. If you envision becoming like the TV and movie version of the scientist in his secret headquarters, controlling happenings in all parts of the world by the push of a button, forget it. Real life is a little less exciting.

ELECTRONIC HARASSMENT

First of all, not everyone is vulnerable to electronic harassment. You will have to make an assessment in your own situation. Secondly, an expenditure of both time and money may be necessary on your part. Again, you will have to decide just how important your situation is.

Before we scare you away, we should also state that if you absorb the basic principles given in this book you can be more sensitive to opportunities that present themselves and be able to do a lot of things that cost little in time or money.

We must divide our electronic techniques into two basic areas:

1. Reception of communications to and from our target, whereby we use the knowledge gained to apply a conventional harassment technique. Communications can be radio type or wire type (telephone, intercom, etc.).

2. Transmission of radio frequency signals designed to produce some action by or against our target.

We should perhaps consider a third area and that is using electrical control circuits to our advantage. You will better understand what we mean when you read "The Heat's On" in a later chapter.

We should learn if our target uses radio communications in any form such as:

1. Hobby (CB, ham, short wave listener, etc.).

2. Radio telephone in his or her vehicle.

3. Commercial radio set-up in his or her private business.

4. Works for an employer in a job that utilizes the employer's radio communications (most public service agencies, utility companies, etc.).

Putting Our Ears On

To monitor any radio communications you will need some form of equipment capable of receiving in the required frequency range and transmission mode (AM, FM, SSB). This doesn't have to be anything elaborate or expensive. Used CB sets are "a dime a dozen" and general coverage receivers are available that cover ham, CB and even public service bands. Police band scanners are as commonplace as toasters.

When obtaining any receiving equipment it should be of a tunable or user programmable nature. This is particularly true in the case of police or public service band receivers and scanners. If a scanner is not programmable then a crystal must be purchased for each frequency covered. We needn't dwell on the hazards of being on record as having purchased such a crystal! In fact, the less you can be on record as having purchased anything the better!

As far as transmitting equipment goes it is perhaps best to go only as far as your electronic background will safely let you. If you happen to be a technician, an engineer or an amateur radio operator (commonly known as "ham"), then it's a different ball game. While a technical discussion on transmitters is beyond the scope of this book, we wish to point out that CB radios are almost a "plug it in and talk" device. Also, signal generators and grid dip meters are readily available and you don't need a PhD in electronics to use them either. Wireless mikes also fall into this category.

The transmission of contraband signals within the public service bands (police, fire, aircraft, etc.) besides being very illegal could endanger countless innocent lives who depend on fast and accurate communication. *Don't do it!*

Enough said on basics and equipment at this time. So get comfortable and prepare to take an entertaining trip into the world of "Electronic Harassment"!

The Honeymoon Sweet

The following is a true story and was done more as a practical joke than a revenge seeking harassment. The "hero" of this story was a young fellow named Ernest. Ernest and his wife were students at a small Midwestern college, where he was a science major and worked part time in a radio and TV service shop. Ernest's wife worked various odd jobs to

help keep bread on the table. Because of work and study, sleep was sometimes a precious commodity for both of them.

Their living quarters consisted of a single bedroom apartment in one of the quadplex units the college had obtained from World War II surplus and provided for married students. Former G.I.'s of that vintage may remember these functional, but basic housing units. Bedrooms were overlapping, with privacy provided by two sheets of thin plaster board. Insulation, apparently, was not an architectural requirement in those trying years.

Ernest's neighbors, in the next apartment, happened to be a newly married couple by the names of Dudley Joe and Theora. They were a friendly young couple, existing on a mixture of poverty and love. For whatever reason, Theora was a compulsive giggler. If you said "good morning" she would respond with a friendly "good morning" followed by a giggle. If asked the time of day, her response would be followed by a giggle.

On this particular Sunday afternoon Ernest was endeavoring to recapture a little lost sleep when he was awakened by sounds coming from Dudley Joe and Theora's bedroom. It was sounds that had awakened him before — muffled conversation, rhythmic bed squeaking, radio playing, and the ever present giggling! On the first occasion, Ernest and his wife found amusement, remembering when they too were newly weds. However, this marathon love making was fast becoming more of a disturbance than an amusement for them.

Radio! Why hadn't he thought of it before? Hurriedly, Ernest brought his newly completed Heathkit SG-8 signal generator into their bedroom. (A signal generator is an electronic servicing device that emits a weak but variable frequency rf signal). With it plugged in and warming up, Ernest fastened the test lead to a short length of hook-up wire. This would serve as a makeshift antenna. The job at hand was to get an rf signal to zero "beat" the radio station that Dudley had tuned in. As the dial of the signal generator was moved across the AM broadcast band a loud whistling sound was heard, nearly covering up the music the honeymooning couple were listening to.

As soon as he was on their frequency Ernest began rocking the dial of the signal generator back and forth in time with the bed squeaking!!! As soon as the squeaking stopped he moved slightly off frequency and the whistling stopped. The bed started squeaking again and so did the synchronized whistle from Dudley's radio! Poor Dudley!!!

Not beaten yet, Dudley changed to another station only to have the "whistle" appear again, in time with the bed as the squeaking resumed. Dudley tried one more station before turning the radio off. By this time his attention had been so diverted by the "phenomena" that any further amorous endeavors were either curtailed or sufficiently quieted.

With one more "treatment" a few weeks later Ernest had successfully used his knowledge of electronics to solve what had been a sticky problem. We might add that another important ingredient of any successful electronic harassment was present in this true story and that was the fact that Dudley had no knowledge of radio or electronics. If so he might have quickly realized what was going on.

The preceding true story is but one way that a signal generator, grid dip meter, or FM wireless microphone could be used to mount a counter-attack against a neighbor's noisy radio or hi-fi receiver playing programs. If they are playing records or tapes then this method can not be used.

Sometimes the proximity of your target's radio will not be as handy as Dudley's was to Ernest. In these cases we must go one step further and that is to effect some type of directivity to the "jamming" signal that we're using. For example let's assume that your target is a next door neighbor in a typical housing development and his TV plays too loud and too late to suit you. A good attack here might be to couple a grid dip meter to a TV antenna. Place this antenna in your attic and aim toward your neighbor's outside TV antenna. If he is using an indoor antenna (rabbit ears, etc.) then aim as close as you can toward it and tune your grid dip meter to the frequency of the station he is watching or listening to in the case of FM stereo. It is best if you can work from a vantage point whereby you can hear when you are on frequency with your "jammer."

That's A Big 10-4, Good Buddy

In classic guerrilla warfare there are definite principles that a wise user will follow with one of these being the "principle of provocation." This is where you are among others who are yet uncommitted toward someone who you wish to make a common enemy. You will cause things to happen to the uncommitted in such a manner that they put the blame on your enemy. In many cases it will be them, not you, that does a number on your targeted enemy. Get the picture?

While many things of a physical harassment nature could be done, the following story illustrates how electronic harassment was used to implement the "principle of provocation".

The "good guy" in this story was a fellow named Willard who worked as a technician for a large electronics firm. Basically a good-natured sort, Willard seemed to have no enemies. However, there was a neighbor across the street and down two houses whom Willard had quietly sworn vengeance upon nearly a year before. This particular neighbor was an active CB'er and had invested in a large amount of CB equipment, all coupled to a large antenna on the roof of his house. Most of the neighborhood was aware of his hobby and "handle", Danny Boy, which was painted on the rear of his camper.

Danny Boy worked for the same firm that Willard did — in fact they even car pooled together. It seems that a year earlier their firm had been expecting a layoff and Danny Boy had tried to undermine Willard's job in order to sacrifice him and save himself. Willard found out about it but chose to let Danny Boy think otherwise.

Willard acquired a CB radio in a swap meet trade and was listening to it one evening when Danny Boy came booming in on one of the channels. Suddenly, Willard had an idea! For the next few months he tape recorded a lot of conversations that Danny Boy had over the CB. By car pooling together he also had the opportunity to secretly record more of good old Danny Boy. It was during these car pool sessions that he really incriminated himself on Willard's secret tape recordings. Danny Boy's next door neighbor was a big palooka who had a very good looking wife. At the slightest provocation, Danny Boy would expound about her and what a shame she couldn't enjoy a stud like him rather than that big jerk husband of her's. Willard would smile to himself as his small pocket tape recorder captured these jewels of incrimination.

When his recordings were sufficient, he began the painstaking chore of editing bits and pieces of various recordings until several simulated CB type conversations were assembled.

A lot of them were laced with Danny Boy's typically off color remarks about his neighbor's wife. By this time you may already have guessed the outcome of this story! Willard built a linear amplifier for his CB set that was sure to get into a few TV sets up and down the block — especially when fed into a folded dipole antenna secreted away in his attic. The hard part was over — now all Willard had to do was wait until strategic times to do a little contraband transmitting with good old Danny Boy getting the full blame when TV sets were affected, which was quite a bit of the time.

The climax of Willard's electronic harassment came when Danny Boy's neighbor punched his lights out after he heard (over his own TV set) the off color remarks about his wife. With the addition of a few more conventional harassment stunts both at home and at work, Danny Boy's life began to crumble with his being none the wiser that Willard, the "friend" whose job he had tried to sabotage, was quietly behind it all.

Many variations of the preceding story could be used. The primary principle was using electronic harassment not directly against your target but against one or more third parties with your target taking the blame.

KITE +
STRING +
KEY +
LIGHTNING
=HOT SEAT!

The Old Repair Call Gimmick

Joe Brown had been considering adding a small computer to his business for some while. He had made enough inquiries for the word to reach the ears of one of the salesmen for central business equipment, better known around town as CBE. Such salesmen proceed to their victim as does a coyote to a wounded rabbit and Joe Brown was no different. To make a long story shorter, Joe was fast talked into a computer setup that cost twice as much as other makes that were available. On top of that he was loaded up with a lot of software and an overpriced service contract and "naturally" would want the special CBE supplies contract.

Joe Brown was no mental match for this super salesman who used every psyschological trick in the book to get Joe's signature on the dotted line or, in this case, lines. By the time the equipment was delivered, Joe realized he had been had but chose to make the best of a bad situation and try to use the new computer the best he could, which was far from what he really needed. Every month, when lease payment time rolled around, Joe rekindled his growing hatred for CBE.

Joe felt helpless against this giant of a company until one day he was listening to an all band receiver at the local radio supply store. While tuning in the neighborhood of 155 MHz on the public service band he heard a strong signal giving what sounded like a repair order. He inquired of the clerk what that signal was. The clerk informed him that was CBE dispatching a repairman to some business. He listened to the next dispatch more carefully — it went something like this: "Number 41 — go to the Miracle Wind-up Radio Company on 3rd Ave., — their paper stretcher is down — contact Shelley Jones at 123-4567." He listened for a few more minutes to the calls going out in a continuous string. Joe lived in a relatively large city and the CBE company was also large.

When the idea struck him, Joe was nearly overcome with excitement and was shaking when he signed the credit card slip. The radio cost nearly a hundred and fifty dollars and he worried about what his wife would say when she saw the bill. His planned use for it was, to him, so sinister that he didn't want her to know the real reason he bought it.

Joe waited until the next afternoon when his part-time help went home and he was alone to start the neat little electronic harassment he had conceived. The radio was tuned to the right frequency after he removed it from the package and turned it on. Sure enough — there was good old CBE dispatching repairmen all over town! Joe still couldn't believe it. Here was his enemy giving him all the information he needed to really raise some hell with them and their customers. Some of their customers naturally were large companies that must purchase thousands of dollars worth of equipment each year. Joe locked the front door, put out the "Closed — Will Return At —" sign, and turned off the lights in the front of his office. He didn't want any disturbances for awhile.

With radio and note pad at hand he made a quick call to the switch board of the CBE company and made an inquiry as to the name of the manager of their service operation. The operator gave him the name of Stan Ettinger, the man who headed their entire service network. Now Joe was ready to get a little revenge -- how sweet it was!

The next dispatch call was for the Wind-up Radio Company again. Their paper clip counting machine was double counting. Please contact Barbara Smith at 123-4567, extension 890. A friendly voice answered "Accounting Department — Barbara Smith." Joe's pulse rate climbed as he began "Barbara, this is Stan Ettinger — service manager for CBE and it just came to my attention that you are having trouble with one of our paper clip counters." This got Barbara's attention. She was pleased that someone so high up at CBE would care enough about her problems to call her personally. Her pleasure was short lived. Joe continued: "our company designs, builds and sells some of the world's finest office equipment and it really dings me off when some klutzy broad such as yourself can't learn to use it properly without hosing something up on it." Barbara nearly dropped the phone and started to reply "Well — I — Who is..." when Joe started in again. "Please don't interrupt me. You have inconvenienced our company enough already by tying up a repairman with another one of your petty problems. A lot of you dingy broads bug us for service when all that's wrong is you forgot to plug it in. Now, please think twice before you waste our time again. Goodbye."

Joe hung up before poor Barbara could catch her breath and respond. He could picture her rushing into her boss's office in tears telling him about the call from the CBE service head. She would be so upset she'd probably add a few things he didn't even say. Joe also thought of the poor, unsuspecting repairman arriving on the scene soon. Oh, Mercy! Well that's what he gets, working for such a bunch of SOB's as CBE.

Joe got so involved in his little project that the afternoon went fast. He decided to hit hard and fast, then back off for a few days. He was smart enough to realize that as soon as CBE figured out what was going on they would contact the police, phone company and Heaven knows who else. It would be a simple matter to set up a few service calls, with the phone number given being one the phone company could have a trace all ready to go on. Joe didn't particularly love Ma Bell and anything he could do to make their life miserable was a welcomed side effect.

On one service call Joe called the number given by the CBE dispatcher and told the lady answering that he was Captain Nelson, the head of CBE security and he would like to have her assistance in a very important matter. "It has been reported to us that one of our service personnel has been making improper advances toward some of the women in offices that he calls on," Joe told her in a most official sounding voice. "We have undercover security people following and observing this individual and we would like to ask you, if you could, to be super friendly to our repairman when he gets there," Joe said. And he continued: "he's not a criminal, ma'am, but, frankly speaking, just too horny to work in this capacity and we must get evidence before our company disciplines him."

Joe was surprised that the gal on the phone not only believed him, but even seemed excited about it. Joe nearly dropped the phone when she offered to have some of the other girls in the office help out. She informed him that one of the girls, besides being a real practical joker, was a real foxxy lady. If she couldn't get this guy to misbehave, no one could. Joe thanked her for her help and assured her that CBE would never forget that she helped them. He even mentioned that, while he couldn't guarantee it in this case, CBE had given new electric typewriters to people who helped them in various ways. He told her how wonderful a company

ELECTRONIC HARASSMENT

CBE was and that he would recommend her to receive a typewriter if she could use one of her own.

The lady was beside herself. She had swallowed hook, line, sinker, rod, reel and half the boat! After Joe hung up he sat back and built a mental picture of what could happen when the poor, unsuspecting serviceman showed up. At first he felt sorry for him then jealous when he thought of the gals flaunting themselves at him.

Joe felt some feelings for the innocent repairmen that were caught up in this little war. However, he thought, in real war it's always the foot soldier that represents the greatest casualties, not the generals. If only he could get to the generals! Well — maybe someday.

This type of electronic harassment could take many forms besides what has been covered in the preceding story. Any business that uses radio dispatching could be vulnerable to having their transmissions intercepted and diversionary action taken. A different twist is presented in the following situation.

May I See Your Driver's License, Please?

John Slater had been a reserve officer in Quietville until he could no longer be a part of what is too typical in small town police departments. Ordinary citizens were harassed for minor things while city officials and prominent businessmen had *carte blanche* to do as they pleased. He had witnessed the city attorney drive through a red light while too drunk to walk. He was merely told to go on home and sleep it off. The same night another less influential person received a citation for throwing a gum wrapper from a car window — a violation of Quietville's new anti-littering ordinance! Needless to say, street crime continued to flourish.

Chief McSwine had only one redeeming quality that insured his job and that was doing exactly what the city fathers wanted, which was not always for the betterment of the ordinary citizen. The balance of the regular officers were typical zombies who had to obey orders or not have a job. John could not exist in such an environment — he hated it and dreamed of ways to fight little bureaucratic dictatorships such as this. John realized his feelings were not popular ones and kept them to himself. In fact, his car sported a "Support Your Local Police" bumper sticker. The ultimate in hypocrisy, he thought!

John had purchased a police band scanner while yet a reservist. The police department used only one frequency for both transmitting and receiving, making what he was about to do easier.

When a police car made a traffic stop, the officer would radio the license number of the stopped vehicle and the physical location by nearest streets. The dispatcher would 10-4 and all would be quiet until the officer went back on the air, indicating that the traffic stop was over. If there was any suspicion on the part of the stopping officer he would ask for the 10-28 ("registered to" information) of the license number and wants and warrants on the driver.

Having been a reserve officer, John was a jump ahead of the average citizen who might attempt to do what he was about to do. John knew the town's troublemakers — who was doing what to who and who wasn't supposed to know about it. He carefully made a card file of every license plate he could that might serve his purpose later. He even thought of buying one of the micro computers he'd seen advertised but couldn't justify the expense. After all, he couldn't appropriate public funds like McSwine could. All in all John probably had a card on half the license plates in town. Every time he heard a license plate mentioned over the scanner he would check his file. If there was a card he would make a notation on it of the date, time and nature of the dispatch. On cards where there were two or more entries he placed a small red marker tab.

John was self-employed and could set his hours within certain limits. He installed another scanner in an attaché case that could lie on the front seat of his car. He chose against a permanent installation for obvious reasons. He was finally ready to begin his electronic harassment project. He felt suddenly powerful — rather like James Bond and Captain Good all rolled into one. He smiled when he thought of what he intended to do to Chief McSwine and the fact that McSwine would be powerless to defend himself.

The scanner's squelch was broken by "Beat One, 908 at Tenth Street and Walnut. Nora Henry John Niner Five Niner — possible deuce." The dispatcher acknowledged, then all was quiet. Being familiar with police codes and jargon, John knew immediately that this was a traffic stop, the location, the license number of the vehicle plus the fact that the driver was probably drunk. He also recognized the license plate without looking it up in his file cards. It belonged to one of Quietville's more colorful characters, Robert Smith. John knew that Smith was less than an ideal husband to a trusting wife, Shirley.

John thumbed through the phone book until he found Smith's number, composed himself briefly, then punched in the number. After three rings, a soft voice answered, "Hello."

John replied, "Am I speaking to Shirley Smith?" "Yes." "Well, Shirley, what I'm about to say to you is a rather rare thing for a Chief of Police to do, but here goes..."

"You mean you're Chief McSwine?" was Shirley's puzzled answer. Just as John was assuring her he was, his scanner came to life. At first it startled him until he suddenly realized that this "official" background noise only helped to prove to Shirley Smith that she was actually talking to Quietville's own Chief McSwine.

"Shirley," John continued, "at this very moment one of my officers has that slob husband of yours stopped for possibly driving while intoxicated. I'm sure you already know about his drinking problem, but are you aware of the other women in his life?"

"Chief McSwine!" she replied in amazement.

John interrupted, "Shirley — — — I don't know how to say this other than just blurt it out — — — Shirley, I'm in love with you. Now I know that this may come as a shock to you, but I think you are a beautiful and kind woman and I can't stand to think of you going to bed with that animal you call a husband. Please don't tell anyone that I called you. It could cost me my job. Just understand that I love you and want to take you away from all the misery that you live with."

Shirley Smith was so shocked that she could only respond with, "I — I — But Chief Mc — I — Oh, my!"

John continued, "Remember, this is just between us. I'll phone you later. Goodnight, Shirley. I love you." John promptly hung up.

John Slater could hardly contain himself as he thought about what was sure to happen. Although Shirley was married to a shnook, she loved and trusted him. The first thing she would do when he got home would be to ascertain if he had been stopped by the police and then proceed to tell him about the phone call she had received from Chief McSwine. John decided to add one final touch to this diabolical little plot. He phoned Shirley's sister, identified himself as Chief McSwine, and proceeded to tell her how he felt about her sister — did she think Shirley would leave her husband for him — and would she please try to persuade her to do so? She too was too shocked to properly respond before John said "Goodby" and hung up. John knew that Shirley's sister was one of Quietville's biggest gossips and the sun would hardly be up before the rumor mill would be grinding away. Poor Old McSwine!

One doesn't need a great imagination to guess what happened to McSwine. With the help of a few more little stunts by John and his scanner, McSwine was almost a basket case. The town fathers no longer wanted a stooge so riddled by scandal and rumor. Soon the city council politely requested his resignation. John knew full well that his replacement would likely be a similar type but at least it would be a change.

John put his scanner away for a while — at least until it might be needed again and turned his attention to a matter that strays somewhat from electronic harassment but was so diabolically clever that it needs to be shared. John never cared for the 55 MPH speed limit — especially the fact that the federal parasites had crammed it down the throats of the nation by blackmailing states that were lax at enforcing it. He knew that what he was about to do would never really change the overall picture. However, he just had to do something to show defiance.

The Highway Patrol had several aircraft that were used to measure automobile speed throughout the state. At selected spots marks were painted on the highway at a specific distance apart. An observer in an aircraft would measure the time it took an automobile to travel the distance between two markers. This time was entered into a small specialized computer which instantly showed the exact speed of the vehicle in miles per hour. Offenders were apprehended by a patrol car that had been alerted by radio.

John studied various schemes to foul what he considered a complete waste of taxpayer money. He knew that crimes of violence against innocent people were increasing daily yet police were busying themselves by using nearly a hundred thousand dollars' worth of equipment plus several men to catch a few drivers going over 55. What a rip-off! He thought of several methods to jam their radio communications but lacked the time or finances to do a good job of this. The aircraft was vulnerable to a lot of ideas but all seemed a little too drastic for him.

He finally decided to do something so simple he had previously overlooked it — repaint their measuring marks. At first thought he was going to extend one of the marks a proper number of feet so a vehicle could be going 75 and yet the spotting aircraft would measure it as 55. He suddenly got a more tactful idea — move the marks closer together so a lot of people would get traffic citations when they were actually within the legal speed limit. Remember the principal of provocation mentioned earlier in this book? If enough innocent people were stopped by the police then maybe enough protests and complaints would be received by the state government that this waste of tax money would be curtailed.

We never learned how successful John's little repaint project was but our imagination can conjure up a lot of interesting scenarios.

Cleared To Land

As this book was being written an interesting act of electronic harassment happened. All readers surely remember the famous air traffic controllers' strike during the summer of 1981. Some may remember reading about false radio messages being sent to airplanes flying under IFR conditions. These messages were made to appear to be coming from official air traffic controllers. Normal air traffic messages give instruction to the pilot as to course bearing changes, altitude changes and other important information. The bottom line is to prevent midair collisions and other hazards that might occur. Since hundreds of innocent lives are involved, only a homicidal maniac would intentionaly interfere with them, let alone transmit false information. Yet this was actually done.

We are not going to give any more technical details of how easy this could be accomplished although anyone who is a pilot will understand!

Naturally, the union denied doing this. (Remember unions always operate within the framework of the law and never do any acts of violence!) And if you believe that you probably believe in the tooth fairy.

However, in all fairness to the union, this could have been done as a "black propaganda" stunt to make the union look bad. In any case this was electronic harassment taken to a deadly extreme to which no rational mind can justify.

The Heat's On

Lester Evans, his wife and two small children lived in an upstairs apartment. This apartment was typical of so many — an old two story house that had been converted into four apartments, two on each floor. The heating was original equipment — a large, gas converted, forced hot air furnace in the basement. The problem was that there was only one thermostat controlling the furnace. As you might have guessed, it was located in one of the downstairs apartments — the one directly over the furnace! To compound an already bad situation, the wife in that apartment happened to be the lady who invented hot flashes, or at least that's the way it appeared to Lester Evans.

During the winter months old Hot Flash kept the thermostat set to what was comfortable for her and to hell with anyone else. Needless to say, the upstairs apartments were like walk-in meat coolers. Polite suggestions to the lady were ineffective. The apartment owner, Sol Weinstein, could have cared less. Lester became determined to be warm for a change even if it meant burning the damn building down!

He went to the basement late at night and studied the thermostat wiring. It was simple — just two wires leading from the furnace control, across the basement to a point where they went up, apparently to the thermostat in Hot Flash's apartment. The thermostat activated the furnace by merely shorting these two wires together. Lester had two choices. One was to simply remove a small bit of insulation from the wires and touch them together. This would bypass the thermostat and turn the furnace on. If carefully done, it would be almost impossible to detect. The problem was that he would have to go to the basement to turn the furnace off and on. The second choice seemed better. The thermostat wires ran behind a telephone junction box before exiting up through the floor. Lester didn't have a telephone although wires and a connector were in his apartment, apparently used by a former tenant.

What Lester did was to carefully unhook the four wires that went from his apartment to the telephone junction box and feed them through a small hole already in the back of the box. Luck was with him because they exited within six inches of the thermostat wires. Next he removed a bit of insulation and connected two of his telephone wires to them, noting the color code of the wires. With the connection taped and carefully pushed behind the telephone junction box, it would have taken a concentrated effort to have discovered his little trick. Now to turn the furnace on all he had to do

was to connect a short jumper wire between two terminals of the telephone connector box in his apartment.

For the first night in weeks Lester's family was comfortable. It was a different story for the Hot Flash family. Before breakfast the next morning, her husband was knocking on Lester's door. When Lester answered the door, it was hard to keep a straight face at what he saw. This poor slob was ringing wet with sweat and greatly concerned about the fact that their thermostat seemed to have run amuck. He wondered if Lester's apartment was also an inferno. Lester said no, that they were quite comfortable and expressed surprise at the thermostat problem. He even offered to accompany Mr. Hot Flash to the basement to check the furnace out. Before going Lester looked back and gave his wife a wink. She knew to unhook the jumper wire as soon as the door closed. When Lester and friend got to the furnace it had shut down and was cooling off.

It only took two more "heat treatments" before Hot Flash got the message and kept the thermostat at a respectful level. In fact, for the rest of the winter, Lester only had to override the thermostat two times, on extremely cold nights. Dangerous, you say? Perhaps so, but so is pneumonia!

You're Bugging Me, Baby!

Electronic bugging, wiretapping, eavesdropping or whatever else you wish to call it has had more than its share of publicity over the years. Beginning with the "Great Seal" incident in our Russian embassy nearly two decades ago, this type of spy activity has been presented in a bad light by legislators (who probably don't want anyone spying on them)

and the media. It is also interesting to note that spying of one form or another has been a standard method of collecting information since Bible times.

Since this type of activity is usually associated with governments and big business, it is an interesting twist when ordinary "little" people use the same methods.

Ed Morris lived in a relatively nice area of town. His duplex was one of several along a two block stretch of Pine Street. When people live in close proximity it is rare when someone is not irritated by something or someone. Most people, fortunately, tend to live and let live. They realize that they too, unknowingly, may do something that bothers another. Ed Morris was that type of guy, except for one thing.

Two duplexes down the street lived two women. They were in their middle twenties, divorced or never married, and each had two small children. The first thing Ed noticed was that their children, all under three, were usually to be found on the sidewalk — dirty, poorly dressed and obviously not watched over by their mothers. Other neighbors noticed this and, also, the presence of a lot of different "uncles" who were always coming and going. It didn't take a genius to figure out what was going on.

Few neighbors, except Ed, seemed to give a damn. Ed probably wouldn't have either except the idea of children growing up in such an environment bothered him. He spoke to a police officer friend of his about the matter. To his amazement, the police not only knew about the gals but treated it more as a joke than anything else. Ed would have forgotten the whole matter except that, because of the proximity of his duplex to theirs, he and his wife began answering knocks on their door only to face some scumbag looking for "Arlene", the name of one of the girls. Ed was ashamed to tell anyone where he lived because by now everyone in town seemed to know about the "red light" area and you can imagine some of the comments he got.

Around lunch time on weekdays, it was not uncommon for a car load of high school boys to spend their lunch hour at Arlene's. Who said that kids waste lunch money on junk food!!! Ed was determined to do something to at least deter the seeming ease with which a prostitution operation could exist with the seeming blessing of everyone. He was beginning to think he was the one out of step.

Had Ed and his close friend and neighbor, Jim Kirby, not been into hobby electronics, this story would have ended with Ed moving to another part of town, away from the "Pine Street Hookers". Instead, they decided to make this a special project. It was decided that the customers, rather than the girls, were the most vulnerable and should be the targets of some electronic harassment. Bugging the girls' duplex was, quite simply, the job at hand. This, of course, was highly illegal. But in this case, it was felt the end justified the means.

Obtaining or building the equipment was not the major problem, getting it into the duplex was. There seemed to be someone there most of the time, which eliminated the possibility of a "Watergate" job. Suddenly they had an idea —design something that could be mailed or sent via UPS to one or both of the girls. This item would have to be something that they would keep, preferably in the bedroom. What could do the job? What would a girl keep in her bedroom? Under what pretense would it be shipped to her? A mis-sent article? Something sent on approval? A lot of other ploys came to mind. No, it had to be personal — like a gift from a lover or admirer. There were certainly plenty of those!

For the next few days Ed and Jim browsed through gift catalogs and stores. The item they needed had to meet certain criteria — reasonably small, relatively inexpensive, something normally kept in the bedroom, something that would be kept out in the open and not stuck on a shelf or in a drawer — and last but not least, allow a small FM transmitter to be installed in it.

Stuffed animals came to mind but were ruled out because of the children. This might do the job otherwise but the chances of an expensive piece of bugging gear winding up in a toy box were too great. Ed's wife finally came up with an optimum solution — a jewelry box or chest. Why hadn't they consulted their wives to start with? A visit to a well-stocked department store revealed at least a dozen different jewelry boxes — all the imported type with drawers, doors, lids, mirrors, etc. A couple even had music boxes that played when the lid was lifted. Whichever one they chose, it could not produce any noise. They finally settled on a medium size chest with lots of drawers and compartments. This one even contained batteries that lit several small light bulbs when the lid was raised. It also had several areas where they could install a false bottom or back to conceal the electronics. It was almost as if the designer had had them in mind!

On the way home Jim suggested that they replace the batteries with an AC power supply, not only for the lights but as a perpetual power source for the FM transmitter they would install. While the C size batteries would last quite a while, the chances of the lid being left open and running them down were present, along with the odds of their not being replaced. It was settled — an AC power supply would be used, plus they would make it failsafe with NiCad batteries that would be recharged by the AC. Should the chest never get plugged in, at least the transmitter would function for the life of the batteries. A power outage would not affect it nor would plugging it into a switch-controlled receptacle.

Three nights later, after a lot of midnight oil and coffee, their project was finished. The receiving end of their setup was simply Ed's FM hi-fi receiver. They fashioned a highly directional antenna from an old TV antenna that looked right in place on Ed's roof. Only someone in the know would have realized that it was pointed directly at Arlene's bedroom, which was about 75 feet away. All was ready except delivery of the chest. It was decided that the chest should contain a note from an unidentified gentleman — apparently one who had partaken of available favors and had even fallen for Arlene in a strange sort of way. Naturally, he would be a discreet sort of fellow who would assume that, because of his great charm, she would know who he was without his name being on the gift. A small bottle of some new, exotic perfume was enclosed with a request that she wear it only for him. A final touch was a small heart-shaped necklace that contained a small diamond in the center. It even had a certificate varifying the diamond. In reality, it was an industrial grade diamond with the entire necklace costing less than ten dollars. It was placed in one of the small drawers, along with the corner torn from a 100 dollar bill. A small note was included, indicating that perhaps the purpose of this chest was to contain future gifts.

The batteries were brought to a full charge and, after a final check-out, the chest was replaced in its original box and then cushioned and packed into a larger, corrugated box. It was addressed to Arlene Jones but contained no return address. Arlene's admirer was a very discreet fellow. After weighing it and affixing sufficient postage, Ed and Jim mailed the package from a town about 30 miles away. Now the wait for delivery.

Jim Kirby lived across the street and down the other way from Ed. With binoculars, it was a good vantage point to watch the mail delivery at Arlene's place. For the next couple of days, Jim's wife would observe the mailman as he delivered across the street. Ed's wife stood by on the phone, ready to listen and record on their FM receiver. Ed had set the transmitter to a vacant spot on the FM dial. The chance of someone else dialing across the dial and picking up the bugging was possible though unlikely. Ed and Jim had quite a laugh thinking what some unknown person would think if they heard what might be transmitted from their little setup.

On the third day a package was delivered to Arlene's duplex. The news was relayed to Ed's wife and on went the FM receiver, which was already set on frequency for the bug. Sure enough, sounds of the postman handling the box were being received — quite loud, in fact. The chest was well packed and only muffled sounds were transmitted as the package was handed to whoever was there to receive it. Ed felt it was very important to hear the conversation when the girls opened their little surprise. It would immediately reveal if their scam was working and he knew the likelihood of the girls mentioning names was great and he wanted those names on tape. He was to be pleased on both counts!

From the ripping sounds transmitted when the package was opened, Ed's wife feared for the safety of the delicate transmitter. As soon as the chest was removed from the inner box, she could hear the entire conversation in Arlene's living room as if she were actually there. By this time Jim's wife had arrived at Ed's house and she was equally ecstatic over their apparent success. If only Ed and Jim could be there! Well, at least they could listen to the tape later.

When Ed and Jim first told their wives about their bugging project, the wives were not too thrilled. A wife could hardly be expected to be pleased at the thought of her husband bugging a cat house! However, after convincing them that they were not perverts and explaining the overall game plan, the wives got into the project with as much enthusiasm as the guys. Security was stressed from the very beginning with the wives double cautioned about telling any of their friends.

The conversation between Arlene and her friend Sally indicated that the scam was working. The dumb broads actually believed that some admirer had bestowed this gift on Arlene. A small hitch ensued that hadn't been allowed for. The gift was aimed toward Arlene simply because, at the time, Ed and Jim didn't know the other girl's name; Now Sally was slightly miffed because she didn't get a gift. Oh, well, some days you just can't please everyone. With Ed's tape deck spinning away, Sally chided Arlene for setting her up with the high school boys and drunks while she took care of the guys who could afford gifts. As it turned out, Sally's fussing at Arlene actually helped remove any doubt that Arlene might have had concerning the validity of the gift. After all, she was the world's greatest piece of tail and why shouldn't men send her gifts!

The bugged conversation then turned to both girls trying to second guess just who the mysterious benefactor was. There was so much background noise it was hard to understand them. It was mostly kids hollering and banging things around. Arlene must have noticed the power cord hanging from the back of the chest because she suggested they take it into the bedroom and plug it in. What luck! They apparently shut the door behind them as the kid noises dropped and the conversation was now crystal clear. The

Eau de Polecat

chest was plugged into a well receptacle and Arlene tried the lid with the lights and all. Next the small drawers were opened with the sound nearly ripping Ed's speakers apart. The perfume, necklace and corner of the 100 dollar bill were then found. Now the guessing game got serious. Arlene exclaimed over the quality perfume and the "real" diamond necklace. It was small, but it was a real diamond. After all, as the song goes, "Diamonds are a girl's best friend." The corner of the 100 dollar bill seemed to please Arlene as much as if the entire bill was there. "Dumb broads!" exclaimed Ed's wife. Arlene was convinced hook, line and sinker that she had landed a big spender. The swankiness of the whole production had pulled it off.

For the next few minutes both girls stood near the chest and discussed many names as possible benefactors, including: a lieutenant on the local police force, two city councilmen, a medical doctor and several prominent businessmen, including a minister of one of the town's more liberal churches. They even speculated that it might be some whackoff nobody. But since that did their egos little good, they went back to naming more prominent men. Ed's wife was amazed to hear what she was hearing. Under other circumstances she would have thought the hookers were just name dropping. However, they were in the privacy of their bedroom (so they thought) and were dead serious about figuring out the mystery. If the little transmitter had quit at that moment, they already had enough on tape to make the whole project worthwhile. They were so excited they could hardly wait for Ed and Jim to get home from work. Jim's wife had already called him at work to tell him all was going as planned. He in turn phoned Ed. Needless to say, the rest of the day went slowly for them.

That evening, as you might have guessed, Ed, Jim and their wives had a ball listening to the tape recording, plus some "live transmissions". The project had succeeded beyond their wildest dreams. As Ed wrote down the names mentioned on the tape he couldn't believe the windfall that was theirs. Reverand Elwood, the liberal minister who had championed every left wing cause imaginable, including gun control. Doctor Feldman, a prime mover in the local abortion clinic. Lieutenant Kerns, who was in charge of their police department's juvenile division. A prize catch was Ralph Guymon, an omni-despised department head at the radio manufacturing plant where Jim worked. In that company, most old employees, no matter how useless, became "sacred cows" with lifetime job security. Ralph Guymon was such a person. It was rumored that upper management would have gotten rid of him long ago had it not been for this caste system that prevailed.

The balance of the names consisted of a few run of the mill local businessmen and a couple of unknowns. That night Ed and Jim made up a file folder for each "customer" so they could build an orderly dossier on each. Over the next month or so they would be adding newspaper clippings and notes when any in their files were mentioned. They and their wives monitored the "bug" as much as possible during this time, recording everything worthwhile. Soon they found it necessary to develop an indexing method to aid in retrieving a particular recording later. Ed's reel to reel recorder had a counter, so it was a simple matter for who ever was monitoring to enter the reel and counter number of anything juicy and incriminating in a notebook. Needless to say, they made many entries!

Of all the visitors to Arlene's, Ralph Guymon made the best recordings. He was a big loudmouth whose voice came over the bug with full fidelity, something he would later regret. He was usually half drunk when he got there and talked freely. One fateful visit he started bragging about some of the other women in his life. One happened to be the radio company vice-president's wife. He went into vivid details about how he put the moves on her at the last Christmas party. He must have been close to or directly facing the jewelry chest while talking.

Arlene and Sally had questioned many "suspects" as to whether or not he was the mysterious giver of the jewelry box, etc. An interesting phenomenon developed — since the girls seemed excited and pleased with the mystery gift, several of their "suspects" hinted that they might just be the one. A few even started bringing small gifts of jewelry, gift certificates, etc. Pavlovian conditioning prevailed when the mention of a gift meant a hasty exit to the bedroom to place it in an awaiting drawer of the chest. Ed and Jim laughed a victor's laugh when thinking of the war these slobs were losing to this little Trojan horse.

Many interesting bits of data were collected by their little bugging setup. They were to learn that Lt. Kerns received favors just for keeping the police off their case. They also learned the girls' unlisted phone number when Sally gave it to someone over the phone in their bedroom. They finally had enough recorded and filed away to concentrate on an attack, or attacks, as the case might be. They decided to concentrate on just a few of the biggest and most vulnerable where they could wreak the most havoc. Oh, they didn't entirely neglect the little guys. Like the night they recognized the voice of a local insurance man at Arlene's. Jim called the John's home and when his wife answered, Jim told her he was a friend of her husband and had just spoken to him in a shopping center parking lot. Her husband told him he was on the way to meet with a client who he thought was a big name underworld character of some sort and he somewhat feared for his safety. He asked him to phone his wife and have her phone him at this place at exactly 9:00 o'clock. Jim told her he would have done the chore himself had he not had a business appointment in a few minutes. She was to call this number (Arlene's) and ask for her husband. When he came to the phone she was to ask him the coded phrase, "Did everything come off okay?" If all was well he would respond, "Yes, we got the hospital account in the bag." If he answered any other way, acted nervous, or asked her, "How the hell did you know where I was?" she was to quickly hang up, call the police and tell them that her husband's life was in extreme danger with the possibility of his being shot. The address she was to give the police was 123 Pine Street, Unit B!

The poor wife was so nervous that Jim had to go over it with her again and suggested she write down the coded phrases as it was important not to call the police unless her husband was really in danger. But, if he failed to respond in

the proper manner to hang up right in the middle of whatever he was saying and get the police there pronto. She thanked him and promised to do as directed. After hanging up the pay phone, Jim drove the few blocks to Ed's place. He just had to see the possible action himself. He and Ed had just popped the tops on a couple of beers when they heard the sounds of cars with tires squealing. A few seconds later flashing red lights seemed to illuminate the entire block. Since they had an excuse, they hurried out to the sidewalk to join other spectators already assembled.

With guns drawn, several uniformed officers took up strategic positions while a sergeant brought out a bull horn and started calling the insurance man's name! When the guy's wife phoned the police she was so excited they probably never got it quite straight whether he was the victim or the villain. About that time the poor John came out, shirt half unbuttoned, tie hanging loosely from his neck. He was raging mad and rushed at the sergeant, shaking his fist. Needless to say, in seconds he was face down on the ground, cuffed and then lifted and shoved into the back seat of a nearby patrol car.

About this time, who should come cruising by but Lt. Kerns. The sergeant walked over to his car and they spoke for a minute or so. Then the lieutenant drove slowly off. He would probably be talking to Arlene later to get the real story. Ed and Jim watched him drive by and thought to themselves, "Don't worry, lieutenant. Your time's coming."

It took Ed and Jim nearly two weeks to perfect a game plan for Lt. Kerns that would kill two birds with one stone. Rev. Elwood was the darling of the anti-gun crowd. He had even assembled a display of a few replica handguns and would always show them to any sympathetic audience that had him address their meetings. He was very masterful at inciting a crowd to a hysterical frenzy by showing a replica handgun while vividly describing how it could kill innocent people. Typical of most anti-gun nuts, his hatred was focused at the gun as if it were a living thing.

Lt. Kerns had, also, been on record as believing that only police should own handguns so it would be quite natural for him (actually Jim Kirby) to phone Rev. Elwood. Jim feared that Elwood had, possibly, spoken to Kerns before and might not buy the imitation he was about to give. He decided to help his act by having the sounds from a police scanner in the background to add authenticity to the scam.

When Rev. Elwood answered, Jim said, "Rev. Elwood, this is Lt. Kerns, police department juvenile officer, and I'm having a meeting at my home next Thursday evening. There is a group of parents concerned about the sale of toy handguns by several of our local merchants and I'd like you to be there with your display of replica guns for a little added emphasis."

Elwood was ecstatic. "Why, yes, but first let me check my appointment book," he replied. In a few moments he returned to the phone and informed "Lt. Kerns" that he was indeed free that evening and would be happy to attend. He also informed Kerns that he had just acquired a replica of a German Schmeisser submachine gun for his little show & tell. Elwood was a master of deception with his little "gun show". As he passed out two or three replica guns for the audience to examine, they were also informed that anyone could purchase these "deadly weapons" by mail order. While this was true, he neglected to tell them that the guns were only replicas, not real ones. His audience seldom knew the difference.

Jim's "Lt. Kerns" expressed his appreciation and told the good preacher to be there at 7:30 sharp. After hanging up, Jim had a horrible thought — what if Elwood happened to call Kerns before the meeting? Then he remembered that Kerns had an unlisted number, although he and Ed had it courtesy of the "jewel box". There was still the possibility of Elwood stopping by the police station although, since most of the officers on the force hated his guts, there was little chance of this happening. There was still a lot left to chance on this setup, more than Ed and Jim would have liked.

On Thursday evening they cruised by Lt. Kerns' house about 6:00 PM. His car was in the driveway indicating that he was home. The next stop was at a phone booth where Ed dialed Kerns' unlisted number. Kerns' wife answered and asked who was calling when Ed asked to speak to the lieutenant. Ed thought fast and told her, "Officer Dugan", whom he knew to be a uniformed officer. When Kerns came on the line Ed began, "Kerns, you probably don't remember me, but I sure as hell remember you. I had to waste several years of my life in a lousy prison on false charges made by you — you scumbag sack of pig crap. Now it's my turn. You had better start sleeping with one eye open, because you're going to pay for ruining my life."

Kerns tried to respond in his official police voice, "Listen, fellow, you'll never get ...!" Click! Ed had hung up.

A few minutes before Elwood's appointed time of 7:30, Ed and Jim were cruising near Kerns' house. Ed's pickup had a camper shell and they had removed the screen from the sliding window on the right side. They picked a street that Elwood would most likely use to get to Kerns' place. Sure enough, they were not to be disappointed. Elwood passed in front of them at an intersection. They crossed over and turned down the next street parallel to Kerns' street. Two blocks later, they made two more right turns and proceeded toward Kerns' house. Timing was critical! It was almost dark, with just enough twilight left to see Elwood get out of his car, reach back in and grab a large attaché case plus what appeared to be a gun. Then they realized that this must be the replica Schmeisser that Elwood mentioned on the phone. They passed in front of Kerns' house just as Elwood was midway up the walk. Jim, who had positioned himself in the camper shell, let go with a Wham-O slinghsot. The projectile, a clear glass marble, went crashing through Kerns' large picture window. The window breaking in front of him so startled Elwood that he froze, never even noticing the pickup as it smoothly moved down the street and around the corner.

There were conflicting stories about what happened next. It seems that Lt. Kerns, after the threatening phone call, was a little jumpy that night. When the window broke he grabbed a 12 gauge shotgun and ran out a side door and to the front of his house to assess the attack from a vantage point. What was he to see but a man standing midway on his walk and

holding a submachine gun in his hand. Elwood was so startled by everything that he turned and ran down the sidewalk. Kerns rounded the corner of his house at full speed, all prepared to give chase on foot. In the excitement he failed to notice the wheelbarrow in his pathway! His cartwheel fall was broken by his shotgun barrel digging about six inches into the soft dirt of a small flower bed. He stumbled to his feet and continued his pursuit. By this time Elwood had about a 100 yard lead and was running flat out. He didn't know what was happening and wasn't about to stick around to find out. He appeared to be getting away and Kerns made the unwitting mistake to fire a warning shot! When the flash and smoke subsided, there was Kerns, face blackened and bleeding in a few small spots, holding what was left of a shotgun! The barrel was peeled back like a banana to within a few inches of the receiver.

Ed's truck was around the corner and out of sight before they heard what they thought to be a gunshot. Jim, who had climbed through the camper boot, looked at Ed. Suddenly they realized that what was intended to be the ultimate in practical jokes had taken on a more serious twist. Although they had no love for Elwood, they really hadn't intended to ice the creep. Oh, well, Elwood always said firearms were dangerous!

Elwood was tackled by a couple of palookas about two blocks down the street from Kerns' house. He was so out of breath he couldn't even talk. Kerns came running up as soon as he assessed his wounds as less than fatal. His face and hands were blackened, making him look like a minstrel on the way to a theater. Little was spoken that night. Elwood was still panting too hard to talk and Kerns was in no mood to listen.

Naturally, the news media took Elwood's side of the conflict with pages of wonderful things he had done for humanity. At the same time they crucified Kerns. Typical headlines were: "Trigger Happy Cop Runs Down Anti-Gun Minister." Kerns was suspended from the force pending investigation. He knew and his lawyers tried to emphasize that he was set up and acted with reason to believe his life was in danger. But the press was ready to crucify him. He was cleared of charges but his police career was on thin ice.

The icing on the cake was when a reporter from the paper, anonymously, received a tape recording of Kerns at Arlene's! The last heard of him he was selling used cars in another state.

From the things Ed heard on his bugging setup, Arlene and Sally were becoming concerned. It seemed like bad things were beginning to happen to some of their better customers. With Lt. Kerns gone, they had lost their guardian angel with the police department. They discussed keeping a very low profile.

Ed and Jim also decided to slow down a bit after the Elwood incident. It was several weeks before opportunity presented itself again. The society page announced that Dr. Feldman would be speaking before a Planned Motherhood group at the Civic Center in a couple of weeks. This might be just the time to do a number on him! Searching their notebook, they found that they had recorded several "sessions" with the good dcctor at Arlene's. The classic was when Sally apparently became pregnant and had Dr. Feldman verify the fact. He came back a few days later and performed an abortion on her within a few feet of the bugged jewel box. Before he began, Sally told him she thought it only fair to warn him that the baby, whose life he was about to snuff out, could be his very own. He grunted that a fetus was a fetus and proceeded with the same feeling as if he was removing a wart.

Jim felt that if they edited and narrated this recording, it would be great to feed into the PA system when Dr. Feldman spoke to the Planned Motherhood group. Making the tape would be easy compared to hooking it into the PA system. They made a tour of the Civic Center where Feldman would speak. The Civic Center was designed to serve as many interests as possible. There were several meeting rooms available by reservation to any worthwhile group. They observed that the locks were all Weisers, with the interior rooms all served by key-in-the-knob locksets. Such a setup is usually master keyed so the janitor or anyone with a need could enter any locked area with the same master key.

Jim approached the lady at the information desk and told her that he and some associates were organizing a new church and his job was to find temporary meeting quarters until they had their own building. She said that several new churches had used their facilities before. Jim asked to see some of their rooms to see if they could use them. She told them that she couldn't leave the desk but handed them a key and told them to check Room 3 in building B. What luck! The key had code numbers on the bow. On a Weiser this is a direct reading code, requiring no code book to interpret it. Out of her sight, Ed quickly wrote the numbers down.

They took a quick look at Room 3, so at least they could converse about it when turning the key back in. Meanwhile, they used a few minutes to really look the place over. Each meeting room had its own PA system with the amplifier in an adjacent closet. The amplifier was coupled to the ceiling-mounted speakers by being plugged into a wall receptacle. From there the wiring was concealed in the wall and ceiling. Perfect, they thought. No doubt access to the attic area would be found in a janitors' closet, etc. They gave the key back to the receptionist, thanked her and left.

From the code number, Ed was able to make a duplicate key. But what they really needed was a master key that would fit any lock in the building complex. This would require taking a lock apart and "reading" the master pins. At home, Ed pinned a cylinder like the one at the Civic Center. On a second, unobserved trip to Room 3, Ed inserted his new key together with a slim steel "tool" that enabled the key to turn to the left and allow the cylinder to be quickly removed from the knob. He then replaced the cylinder with the one he had pinned at home and double checked that the key would open the door. He slipped the original cylinder into his pocket and back home he went.

By carefully disassembling the cylinder, he was able to observe and measure the additional "master" pins in the cylinder. He wrote these down and put the original cylinder back together. The new measurements allowed him to cut a master key. He tried it in the cylinder and it worked. Now all he had to do was to replace this cylinder in Room 3 the same way he got it. This he accomplished the next day after work. While there he tried his new master key in several other locks and it worked perfectly. This little "Open Sesame" would take him and Jim anywhere in the Civic Center they desired to go. Now back to the electronic part of this project.

While Ed was doing the locksmithing, Jim and his wife had edited and narrated a tape recording to be played to Dr. Feldman's audience. They had first decided to place a cheap tape recorder and amplifier in the attic and activate it remotely. However, their final solution was to interrupt the wires to the ceiling speakers and re-route them to another spot, where they could either be hooked up to function normally or switched to another sound source. The spot chosen was a janitor's closet down the hall. The method to switch to their sound source (a cheap tape recorder and amplifier) would be a clock rigged like in a time bomb.

The meeting was scheduled for 7:00. Jim suggested setting the time for 7:30. Meetings usually start 15 or 20 minutes late. This would allow 10 to 15 minutes for the spokesman for the Planned Motherhood group to introduce Dr. Feldman. Since the meeting had been announced and was open to the public as a fund raiser, Ed and Jim thought it would be nice if they and their wives attended. After all, they had been working pretty hard on the jewel box project and deserved a night out!

They arrived at precisely 7:00 and quietly took seats in the already three-fourths filled auditorium. Ed wondered about their clock ticking away on the shelf of the janitor's closet. Was it still running? Would their amplifier be powerful enough? Were the speakers 8 ohm? If something failed, a lot of preparation and hard work would have been in vain. They were not to be disappointed! Soon Dr. Feldman and some lady, apparently the head of the Planned Motherhood group, took chairs near the speaker's platform. Ed observed that the lady had two distinguishing features. First, she wore too much makeup and second, not near enough to cover up her case of terminal uglies. As he looked around the room at these pro-abortionist fellow travelers he noticed how they had one thing in common. They were all ugly. Too bad their parents didn't practice abortion, he thought.

The chairlady called everyone to order and proceeded to make an introduction for Dr. Feldman. She also introduced his wife, who was sitting on the front row. The introduction was typical — praises for the doctor, what a great humanitarian he was and how he had dedicated his life to helping "girls in trouble". Feldman took the microphone and made a few equally stupid remarks about how great the Planned Motherhood organization was. He took some notes from his pocket and was about to proceed when the microphone went dead. There was a loud crack from the speakers. Ed and Jim braced themselves. Just as Feldman was thumping the microphone in a vain effort to fix it, a female voice came over the speakers loud and clear.

"Dr. Feldman — this is your life" blurted from the speakers along with a few other statements to get everyone's attention. By now Feldman had quit thumping the microphone and was just standing there looking dumb. Apparently he thought the sponsoring group had surprised him with some sort of an award. The lady in charge of the meeting must have thought that this was an attention getting shenanigan of Dr. Feldman's. The voice over the speakers continued, "And now, Dr. Feldman, to present you with your rewards are two very special women in your life — Arlene and Sally." You could have heard a pin drop. Feldman was frozen. His wife just sat there with her mouth hanging open. "Hello Dr. Feldman — this is Arlene and on this occasion I'm happy to recall some of the good times we have had together." Jim's wife had done a superb job of imitating Arlene's voice — at least good enough under the circumstances. She continued, "The good time I remember best there were four of us — me, Sally, you and that doctor friend of yours who was visiting you." Pure fear and panic now were evident by the appearance of Dr. Feldman. He grabbed the microphone again and ripped it loose from the receptacle where it was plugged in and threw it up against the wall behind him. He was becoming a savage animal — wanting to attack something but not knowing what. The voice continued, "Let me refresh your memory, Dr. Feldman. I'm sure you'll remember after this." The next sounds from the speakers was the recording Ed had made when Feldman had treated his visiting colleague to a few thrills at Arlene's. Feldman and his friend were about half sauced that night and wanted a little four-on-the-floor action which Arlene and Sally were happy to provide. There was a lot of conversation and sounds picked up by the bugged jewel box that night — enough that it didn't take much imagination to conjur up a picture of what was taking place.

Feldman was, by now, a raging maniac, His sobbing wife was shoved aside as he threw a folding chair at one of the ceiling speakers. It glanced off the ceiling and landed on a giant of a bald headed man about eight rows back, cutting his forehead and dazing him for a few seconds. When he shook his head to consciousness and saw the blood now staining the front of his shirt, he went berserk and made a lunge at Feldman. By this time, Feldman was choking the lady who had introduced him, blaming her for setting him up like this. The entire room was fast becoming a full-fledged riot, with the speakers continuing to spout forth "this is your life" episodes for Dr. Feldman. Ed, Jim and their wives had decided they had better make a quiet exit like a few others had already done. One of the last things they noticed was one fellow was standing on a table he had pulled over under one of the speakers. Even a riot was not going to stop this guy from hearing this X rated stuff. By this time windows were being smashed and someone must have called the police because Ed's car had to pull over to let the Code Three police car go by.

The alarm clock was set to switch the speakers to the regular PA amplifier after 20 minutes. So by the time the police arrived the PA system would have been quiet. All they had was a bloody mess of a riot on their hands. The big bald headed guy had gotten to Feldman and nearly beat him to death before four men in blue succeeded in pulling him off, only to have him attack them. By this time the four police officers were swinging clubs and spraying Mace.

The TV crew got there just behind the police and were busy filming and interviewing anyone that would stand still. From what the TV and newspaper later reported there was little resemblance to what really happened. Only Ed or Jim knew the truth and they weren't about to tell! From this one little shenanigan would come 7 arrests and 5 civil suits amounting to over 3 million dollars. What a night!

The very next afternoon Feldman was at Arlene's screaming at her for leaking gossip about him. Naturally, she didn't know what he was talking about and told him he was crazy which got her a backhand to the teeth. Ed's wife, who was monitoring on the FM receiver, had phoned the police by now. She didn't identify herself, only telling them that Dr. Feldman, the crazy who had nearly destroyed the Civic Center the night before and was now out on bail, was in the process of assaulting a woman at 123 Pine St., rear duplex. She added emphasis by telling them she feared for the woman's life and thought she just heard shots fired. Of course she didn't but she knew "shots fired" guaranteed fast response from police. Within seconds the street was alive with patrol cars and in a few more minutes, Dr. Feldman was cuffed and shoved into the back seat of a patrol car. Poor Dr. Feldman. Poor Arlene. Trouble everywhere!

By now Feldman had lost what little credibility he had left. There wasn't even any investigation into his claim of being framed. In a couple of weeks Jim was able to retrieve their tape recorder setup from the civic center. He taped the speaker wires together and tucked them under some insulation, all ready to be used again if necessary!

Ralph Guymon's treatment was rather dull when compared to his fellow members of Arlene's fan club. The Christmas season was fast approaching and Guymon's company was preparing for their annual Christmas ball which was usually held on a Friday night and was preceded in the afternoon by an exchange of presents around the office Christmas tree. The whole production would have been boring except for the fact that half the employees were usually smashed by mid-afternoon.

With the success of their previous narrated tape recording behind them, Jim and his wife had little trouble doing one for good old Ralphie. They bought a dozen John Boulder cassette albums while they were on sale. Jim bulk erased them so they could re-record from the master they were making. Pre-recorded tapes have failsafe notches to prevent accidentally recording over what is on the tape. Jim placed a small piece of Scotch tape over the notches to facilitate recording and removed them when done. This recording would be a lot less subtle than Dr. Feldman's since most of the people would be feeling no pain while listening. They did a sort of poor man's version of "Candid Microphone", a radio show that preceded the TV version — "Candid Camera". Jim's wife again did her imitation of Arlene's voice. With a background of Christmas music, Arlene proceeded to tell Santa Claus what she wanted for Christmas. Had she been a good girl? Why, yes — nearly all the men she knew had told her how good she was! The background music diminished and now you were listening to Ralph Guymon's voice telling Arlene how "good" she was. His voice was punctuated by some very suggestive bedroom noises.

The tape continued with Arlene telling Santa that all she wanted for Christmas was Ralph Guymon. In fact, the entire 20 minute tape was of Arlene talking about Ralph. The last episode was her asking Ralph to repeat how much he enjoyed putting the make on the vice-president's wife at the last Christmas ball. Again, there was Ralph's stupid voice "telling it all".

The tapes were carefully gift wrapped and addressed to some "key" people at the company. Naturally, there was no mention of who the giver was. Jim, careful not to be seen, placed the "gifts" under the office tree along with the dozens of others. On the fateful afternoon the presents were distributed and opened in typical office party style. The John Boulder albums would have gotten little attention except for two facts: the giver's name was absent, although this was not that unusual and Jim's wife, after applying a bright red lipstick, had pressed her lips on a small blank card and placed one in each album so the lip prints would show! This should generate a sufficient amount of curiosity, especially among the male gender, they thought.

The tapes were "given" to people who Jim knew would do the most to "spread the word". Two copies went to guys in the model shop area. Jim knew these clowns would probably play one over the PA system if they knew how and were sober enough, which was doubtful. When the presents were opened there was quite a stir over the mysterious John Boulder tapes, especially the fact they all contained a card with a lip print on it. However, only one guy was curious enough to dig up a cassette tape player and listen to his copy. He listened privately first, because he thought it might be something "personal." When finished, he could hardly contain himself. He didn't exactly know why he was given a copy but it was too good not to share with a few close buddies. He quickly rounded up five of them and out they went to the parking lot to one of their cars that had a cassette player. It happened that none of these fellows liked Ralph and the rumor spread fast that everyone should listen to their "John Boulder" album.

Jim went to a car with some friends to listen to their copy. He had to be quite an actor with his reactions while listening. They were nearly through and roaring with laughter, when Ralph was observed walking to his car, obviously to play the tape that Jim had so thoughtfully given him! He no doubt had heard some of the gossip and decided to listen to the tape alone.

He couldn't have been over half through the tape when his car started up and went squealing from the parking lot. Ralph wasn't to be seen the rest of the day and his presence was lacking at the ball that night. In fact, he called in sick after Christmas, apparently using the time off to look for another job. He cleaned out his desk after work one evening and was never seen again. His presence was missed by few.

Meanwhile, back at the "raunch" house Arlene and Sally had decided that a dark cloud had somehow blown over their little enterprise and were in the process of moving away from Pine Street. Ed and Jim were in front of Ed's duplex with the ploy of tuning up Jim's car as the U-Haul was being loaded

ELECTRONIC HARASSMENT

CASHION OR BUST!

in front of Arlene's. Jim nudged Ed to look at Sally carrying the little jewel chest out to the trailer. They were pleased that the little "broadcasting station" would live on at a new location. They could not help laughing when they thought what the reaction would be when some future neighbor of Arlene's would innocently tune across 94.5 on the FM dial!

"You know, Ed," Jim said, "we ought to write a book about all this someday.

"Yeah," replied Ed, "but who the hell would believe it?"

KITE +
STRING +
KEY +
LIGHTNING
=HOT SEAT!